Solar System

SUN AND MOON

Rosalind Mist

QEB Publishing

Words in bold can be found in the glossary on page 22.

Published in the United States by
QEB Publishing, Inc.
23062 La Cadena Drive
Laguna Hills, CA 92653

www.qeb-publishing.com

Library of Congress Control Number:
2008012590

ISBN 978 1 59566 661 1

Printed and bound in China

Author Rosalind Mist
Consultant Terry Jennings
Editor Amanda Askew
Designer Melissa Alaverdy
Picture Researcher Maria Joannou
Illustrator Richard Burgess

Publisher Steve Evans
Creative Director Zeta Davies

Picture credits
(fc=front cover, t=top, b=bottom, l=left, r=right)

Corbis JPL/USGS 1b, NASA 5b

NASA fc,10, 14, 17, 20–21, 20l, 21r, 23b, 24, ESA/SOHO 11, JPL/USGS 15, 18, 23, 24

Shutterstock 1t, 2–3, 4–5, 6–7, 8, 9, 13t, 13c, 13b, 16, 19l, 19r

Contents

The Solar System 4

The Sun 6

Sunlight 8

Stormy Sun 10

Solar eclipse 12

The Moon 14

On the surface 16

Full Moon, Half Moon 18

Flying to the Moon 20

Glossary 22

Index 23

Notes for parents and teachers 24

The Solar System

The Solar System is made up of the Sun, and everything that orbits**, or circles, it.** This includes the planets and their moons, as well as **meteors, asteroids,** and **comets**.

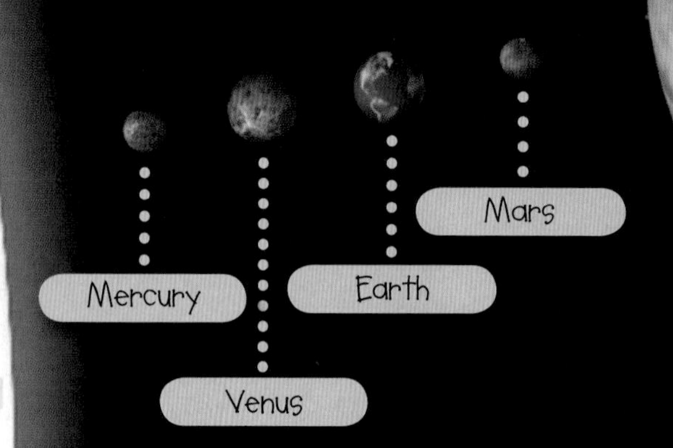

Jupiter

The Sun is at the centre of the Solar System. It is 100 million miles from Earth.

Sun

STAR FACT!
We can see some planets shining in the night sky because they reflect light from the Sun.

**The sizes of the planets are roughly to scale, but the distances between them are not to scale.*

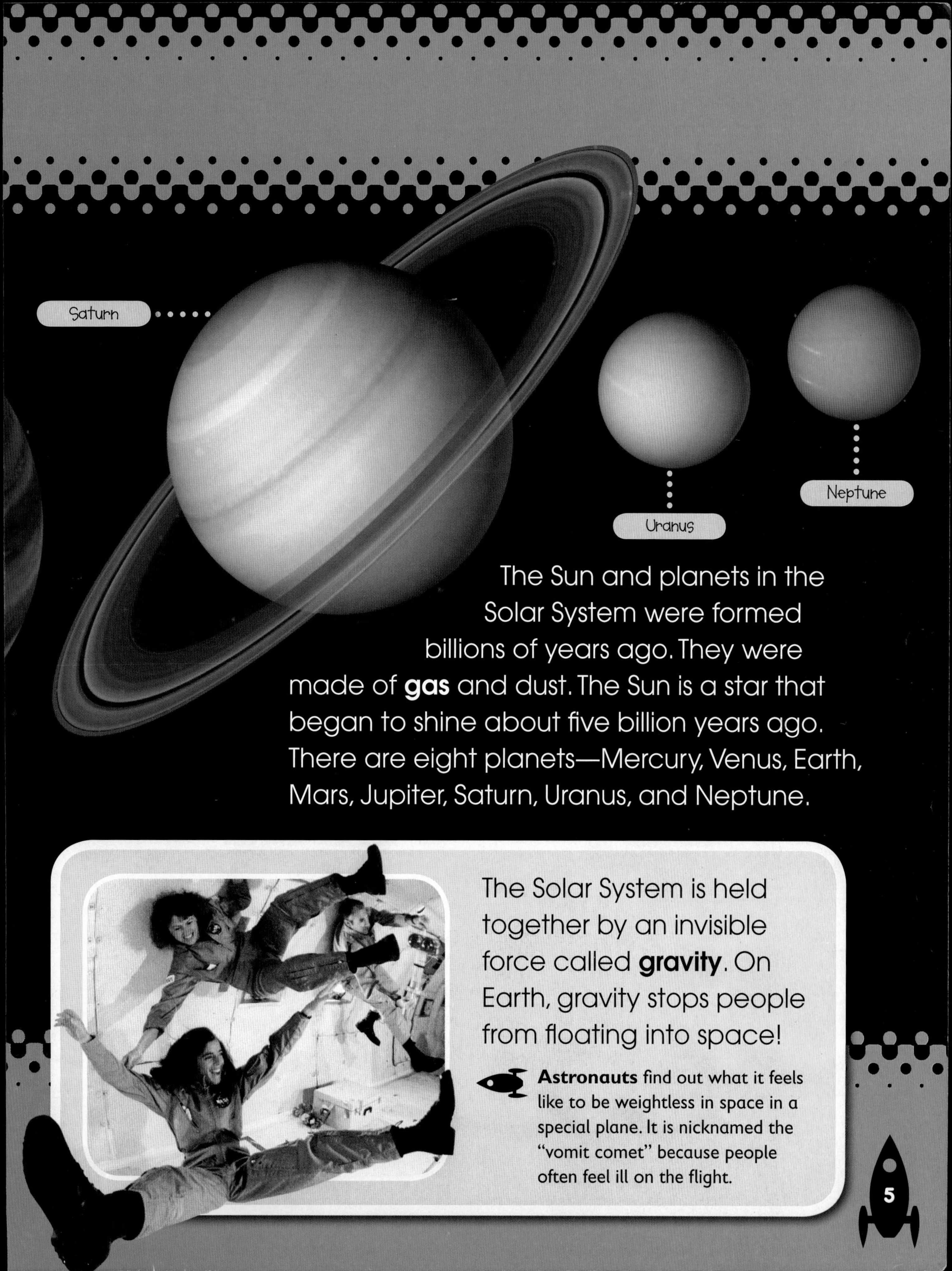

The Sun and planets in the Solar System were formed billions of years ago. They were made of **gas** and dust. The Sun is a star that began to shine about five billion years ago. There are eight planets—Mercury, Venus, Earth, Mars, Jupiter, Saturn, Uranus, and Neptune.

The Solar System is held together by an invisible force called **gravity**. On Earth, gravity stops people from floating into space!

Astronauts find out what it feels like to be weightless in space in a special plane. It is nicknamed the "vomit comet" because people often feel ill on the flight.

The Sun

The Sun is a star, just like the stars that you can see in the sky at night. It is at the center of the Solar System and is the nearest star to the Earth.

The Sun is not solid like Earth, but is a huge ball of very hot, burning gas. It sends out heat and light into space.

STAR FACT!
The Sun looks larger than other stars because it is nearer to the Earth. The next star is nearly 300,000 times further away.

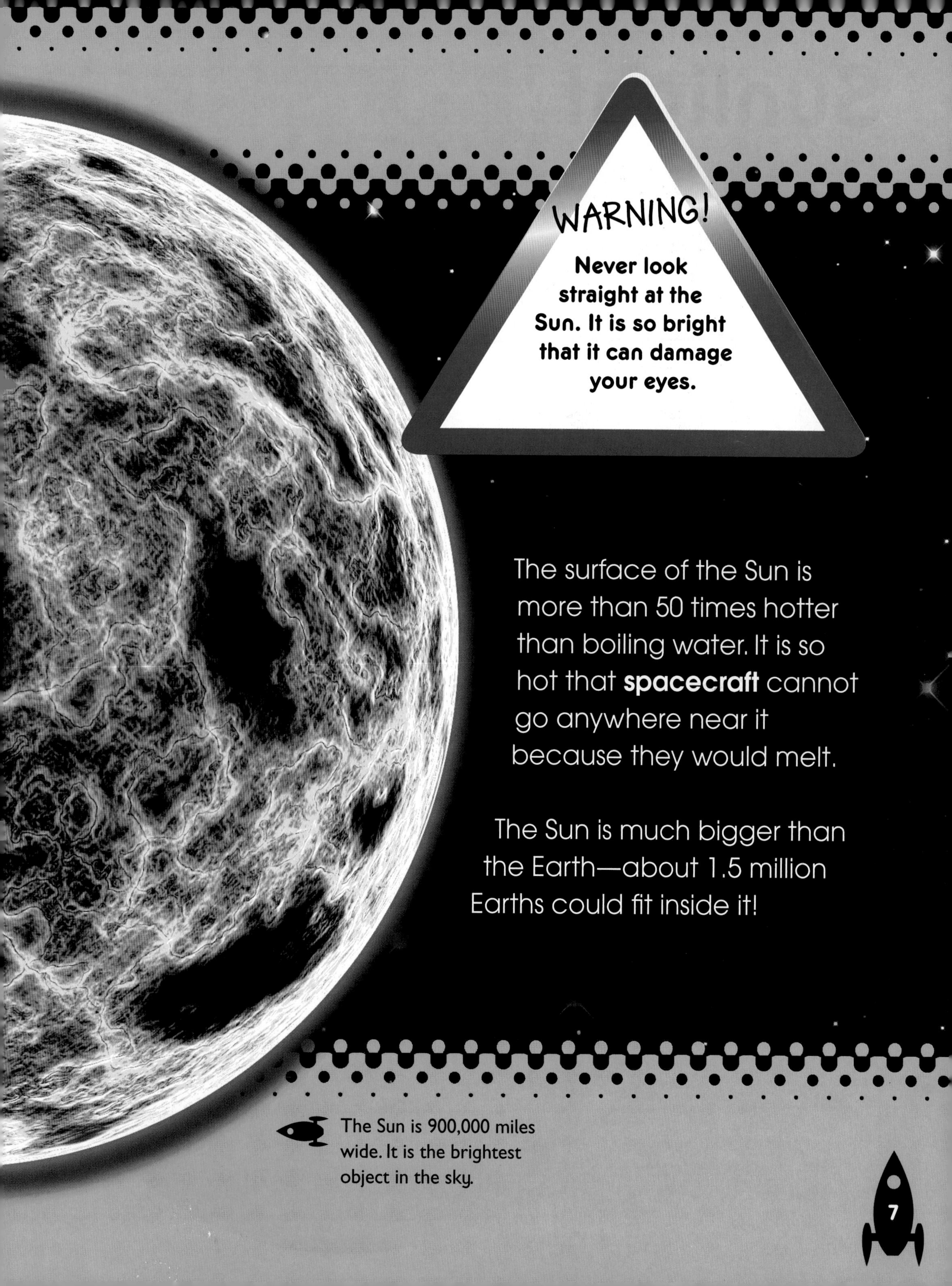

The surface of the Sun is more than 50 times hotter than boiling water. It is so hot that **spacecraft** cannot go anywhere near it because they would melt.

The Sun is much bigger than the Earth—about 1.5 million Earths could fit inside it!

The Sun is 900,000 miles wide. It is the brightest object in the sky.

Sunlight

The Sun produces light, which travels 100 million miles to the Earth.
It only takes eight minutes and 20 seconds for the Sun's rays to reach the Earth.

We can only see things when there is light. At nighttime, without any sunlight, it is dark and colder than in the daytime.

When one side of the Earth faces the Sun, it is daytime. On the opposite side of the Earth, where there is no sunlight, it is nighttime.

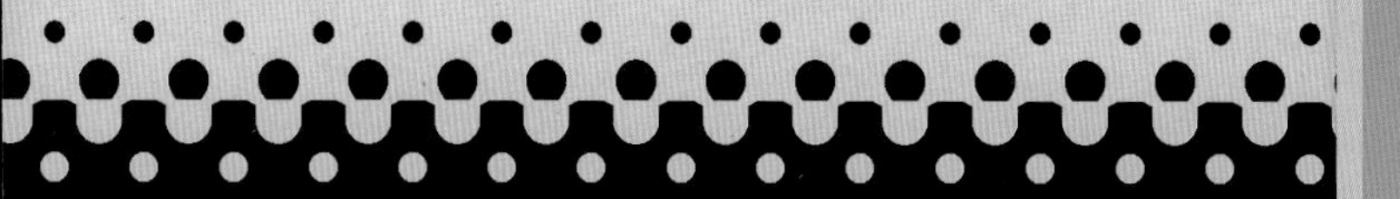

The Sun is very important. Without the Sun, plants would not grow. Animals, including humans, need plants to eat.

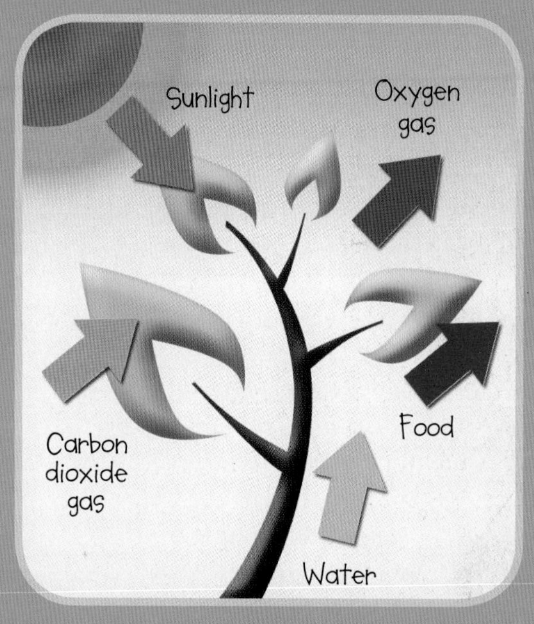

Plants use sunlight to turn **carbon dioxide** gas and water into food and **oxygen** gas.

Sunlight is made of many different colors, not just yellow. There are seven colors—red, orange, yellow, green, blue, indigo, and violet. When sunlight passes through raindrops, a rainbow forms and all the colors can be seen.

A rainbow makes a curve across the sky. The outside edge of a rainbow is red and the inside edge is violet.

Make your own rainbow

Stand next to a wall and shine a torch on an old CD. Move the CD around. Can you see a rainbow on the wall? Can you see the seven different colors?

Stormy Sun

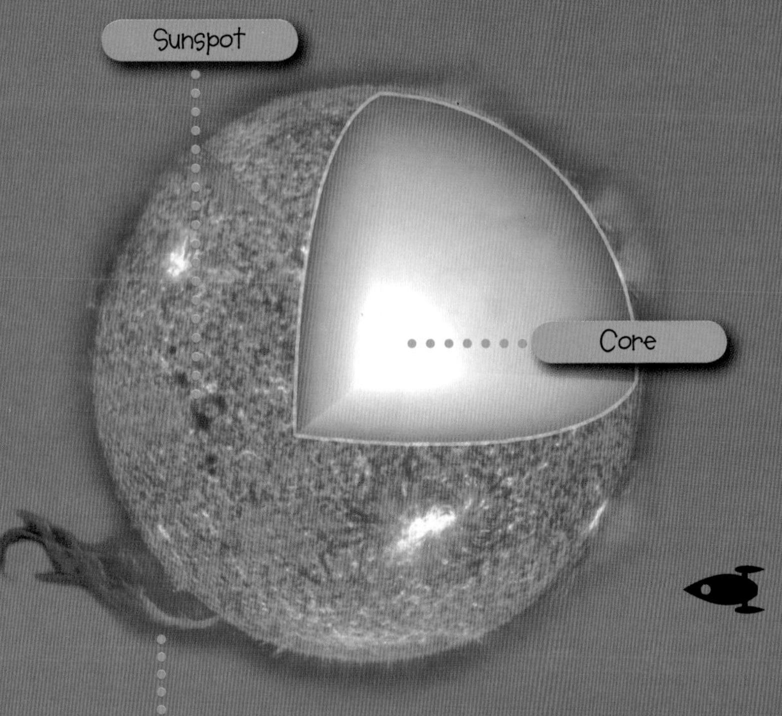

The Sun produces heat as well as light.
The core, or middle, of the Sun is the hottest part. Hot gases bubble up from the core to the surface of the Sun. When the gases burn, heat and light are produced.

The Sun is made of many layers. The core is nearly 3,000 times hotter than the surface.

The surface of the Sun has dark areas called sunspots. They are cooler than the hot, yellow areas.

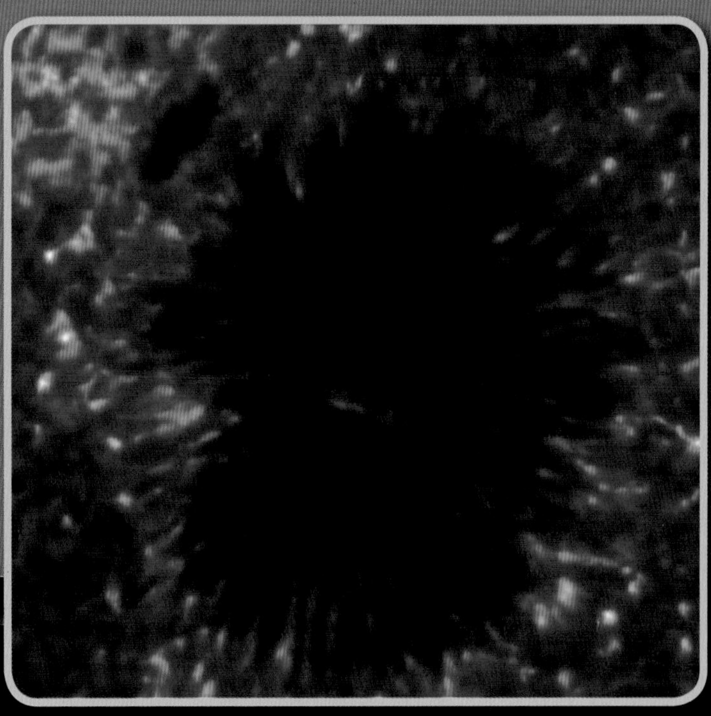

Sunspots were discovered by the **astronomer**, Galileo Galilei. Some sunspots can last for weeks or even months.

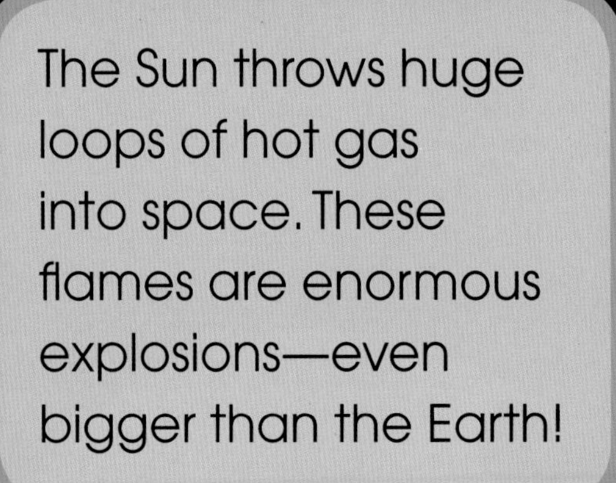
The Sun throws huge loops of hot gas into space. These flames are enormous explosions—even bigger than the Earth!

As the gases on the Sun burn, they create huge fiery storms. This flame of gas is about 190,000 miles across —that is more than 23 Earths!

Flame of gas

STAR FACT!

We can use the Sun's **energy**. Some **solar panels** soak up the Sun's heat to warm water. Others soak up light and then change it into electricity.

Solar eclipse

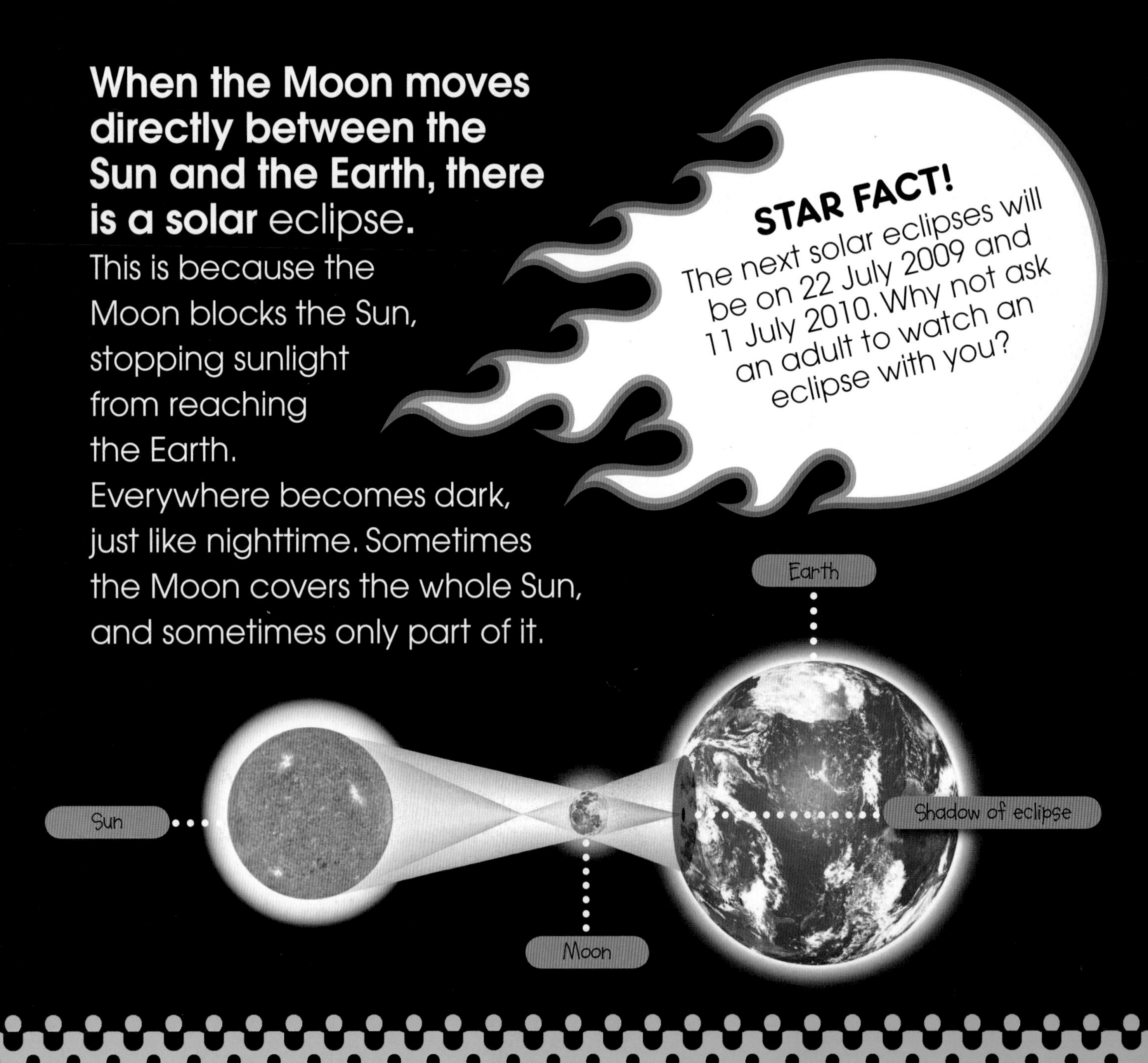

When the Moon moves directly between the Sun and the Earth, there is a solar eclipse.

This is because the Moon blocks the Sun, stopping sunlight from reaching the Earth. Everywhere becomes dark, just like nighttime. Sometimes the Moon covers the whole Sun, and sometimes only part of it.

STAR FACT!

The next solar eclipses will be on 22 July 2009 and 11 July 2010. Why not ask an adult to watch an eclipse with you?

When the Moon completely blocks the Sun, it is called a total eclipse.

Special solar eclipse glasses block out the dangerous rays of sunlight to protect your eyes.

People need to wear special glasses when looking at an eclipse otherwise their eyes could be damaged. Scientists can see the hot gas around the Sun during an eclipse. The gas stretches for 650,000 miles into space.

The hot gas around the Sun is called the **corona** and normally it cannot be seen.

When the Moon first starts to move away, the Sun's corona looks like a sparkling diamond ring.

The Moon

The Moon is a round, hard, rocky ball.

It is smaller than the Earth—in fact, nearly four Moons could fit across the Earth. The Moon orbits, or circles, the Earth.

The Moon is the brightest object in the night sky.

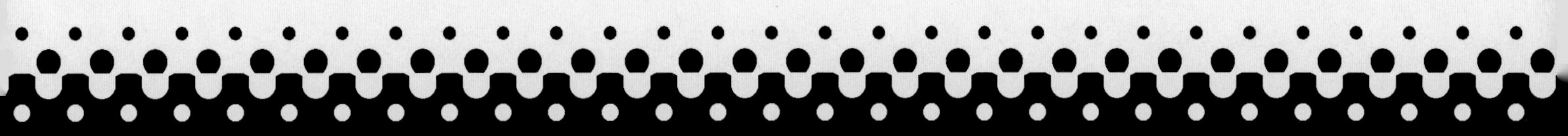

The Moon is made of pale-colored rocks. These rocks reflect the light from the Sun. The Moon does not make its own light.

The Sun and Moon seem to be about the same size from Earth, but really the Sun is much bigger than the Moon. They only seem to be the same size because the Sun is a lot further away than the Moon.

Largest crater

We always see the same side of the Moon. By taking rockets into space, we can see what the other side of the Moon looks like. There are more craters and fewer dark areas.

The largest crater in the Solar System is on the far side of the Moon. It is 1,500 miles across—about half as big as the United States.

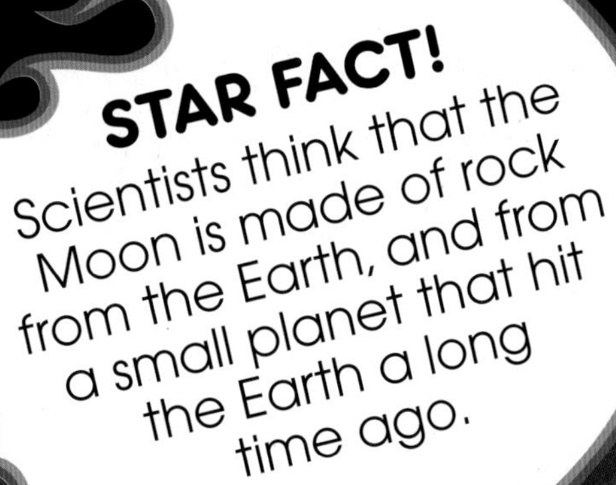

STAR FACT!
Scientists think that the Moon is made of rock from the Earth, and from a small planet that hit the Earth a long time ago.

On the surface

When rocks crash into the Moon, they make holes called craters. Some craters are so wide that a large city, such as London, England, could fit inside! There are about 500,000 craters that are larger than 0.5 miles—that is as long as 20 Olympic swimming pools.

STAR FACT!
You can see the Moon in the daytime as well as at nighttime, but it is often more difficult to find.

The Moon also has lots of dark and light areas. The dark areas are called plains. They are quite flat and filled with rock made from **lava**. The light areas are higher than the plains.

The Moon has many features on its surface. The dark patches are called plains. The light areas are covered in craters.

Moon gazing

With an adult, go outside and look at the Moon. Use a pair of binoculars or a telescope if possible. Can you see the dark areas called plains? Can you see the craters?

Rock and dust around the edge of a crater were thrown out when the crater was made.

Full Moon, Half Moon

It takes about one month for the Moon to go around the Earth once.

The Moon seems to change shape because as it moves, different bits are in sunlight. We see the parts that reflect sunlight. The rest of the Moon is in shadow and we cannot see it.

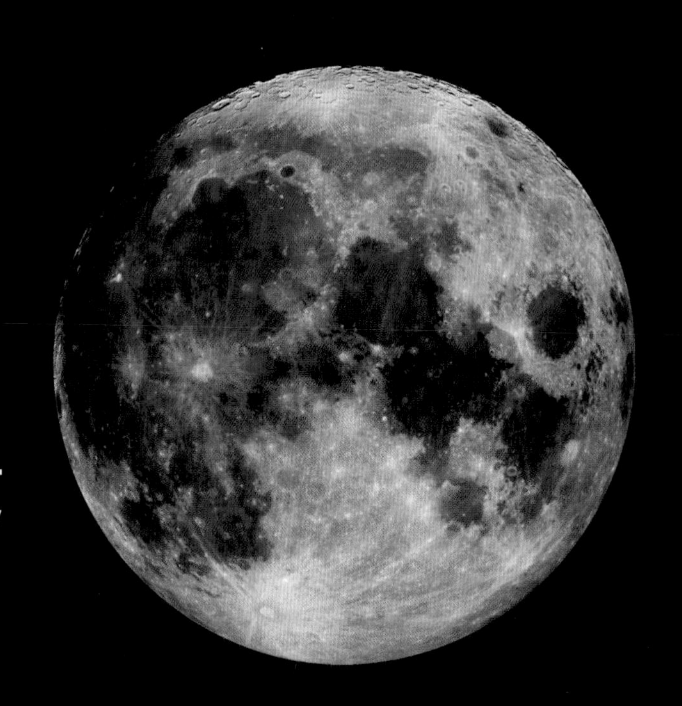

Full Moon

The changing shapes of the bright part of the Moon are called the phases of the Moon.

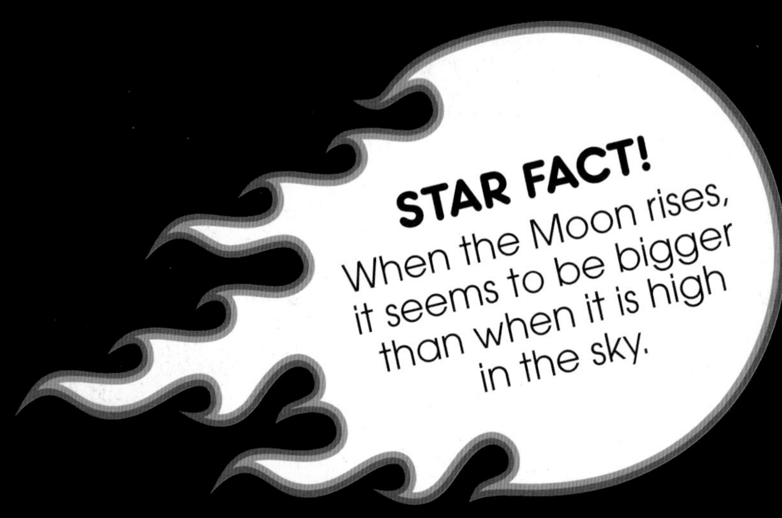

When the Moon looks round, it is called a Full Moon. When only part of the Moon can be seen, it is called a Half Moon.

Half Moon

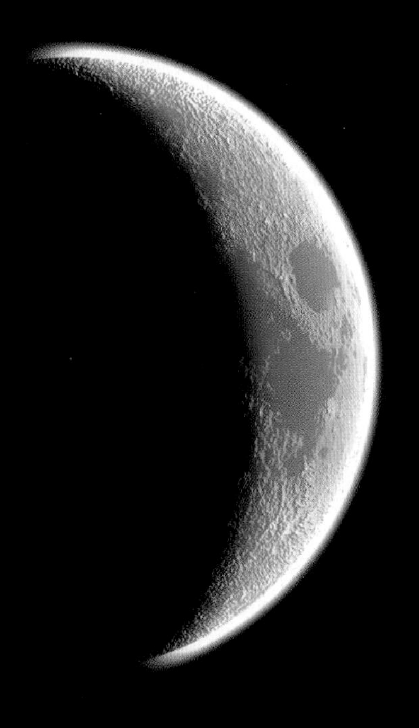

Crescent Moon

A Crescent Moon looks like a big bite has been taken out of it! When there is a New Moon, it is in complete shadow and we cannot see the Moon at all.

Moon study

For a month, draw what the Moon looks like each night. Can you label the different phases of the Moon? Can you explain why the Moon changes shape?

Flying to the Moon

It is a long way to the Moon —about 240,000 miles.

If you could drive a car to the Moon, it would take more than four months to get there! Astronauts are people who go into space. They travel in rockets and shuttles, which move very quickly.

STAR FACT!
On Earth, we see the Moon rise and set. Astronauts who orbit the Moon see the Earth rise and set instead!

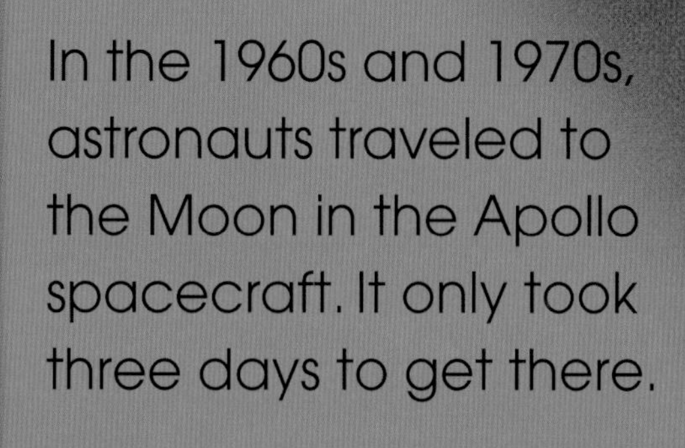

In the 1960s and 1970s, astronauts traveled to the Moon in the Apollo spacecraft. It only took three days to get there.

The most famous mission to the Moon was the *Apollo 11* mission in 1969. Neil Armstrong and Buzz Aldrin were the first humans to walk on the Moon.

Today, astronauts go into space in the **space shuttle.** The shuttle goes to the **International Space Station**, not the Moon.

Astronauts used lunar **rovers** to travel around the Moon. They brought some Moon rocks back to Earth. Scientists used them to find out how old the Moon is.

When astronauts drove around the Moon, they found it was rocky, dusty, and dry.

Scientists also send **space probes** to the Moon. They carry experiments to tell scientists what the Moon is made of. So far, the probes have found cooled lava, metals, and even ice!

Glossary

Asteroid
A large lump of rock, too small to be a planet or dwarf planet.

Astronaut
A person who travels in space.

Astronomer
A scientist who studies the Solar System, stars, and galaxies.

Carbon dioxide
Colorless gas needed by plants to grow.

Comet
An object in space made of rock and ice.

Corona
Hot gas around the Sun.

Eclipse
When the Moon moves between the Earth and the Sun.

Energy
A source of power, such as electricity.

Gas
A substance, such as air, that is not solid or liquid. Gas cannot usually be seen.

Gravity
Attractive pulling force between any massive objects.

International Space Station
Large space laboratory where astronauts can live for months.

Lava
Molten, or liquid, rock that has cooled and turned into a solid.

Meteor
A glowing trail in the sky left by a small piece of rock from space.

Orbit
The path of one body around another, such as a planet around the Sun.

Oxygen
Colorless gas needed by plants and animals to breathe.

Rover
A lunar rover is a car that astronauts drive on the Moon.

Solar panel
A panel that changes the Sun's energy into electricity or heat.

Spacecraft
A vehicle that travels in space.

Space probe
A spacecraft without people on board.

Space shuttle
Spacecraft that has wings to return to Earth like a glider.

Index

astronauts 5, 20–21

corona 13
craters 15, 16–17

day 8, 16

energy 11

Galileo Galilei 10

heat 6, 10

light 4, 6, 8–9, 10, 15, 18

Moon 14–15
- astronauts 20–21
- craters 16–17
- eclipse 12–13
- phases 18–19

moons 4

night 8, 16

plants 9

rainbow 9
rockets 15, 20

Solar System 4–5
solar panels 11
spacecraft 20
space probes 21
space shuttles 20, 21
star 5, 6

Sun 6–7, 15
- eclipse 12–13
- heat 8–9, 10–11
- light 8–9, 10–11
- Solar System 4–5

sunspots 10

Notes for parents and teachers

- Never look at the Sun directly with your eyes, binoculars, or a telescope. The Sun's rays can damage your eyesight. If you decide to look at the Moon during the day, take care that your child does not point binoculars at the Sun. This could damage their eyesight permanently.
- The Sun is much bigger than the Moon. However, if you look at them both in the sky, they seem to be a similar size. This is a perspective effect and happens because the Sun is further away than the Moon. The Moon is 400 times smaller than the Sun, but the Sun is 400 times further away. The average distance to the Moon is 240,000 miles. The Sun is 100 million miles away.
- We can see the Moon because it reflects light from the Sun. It is not a source of light.
- Solar eclipses are caused by the Moon moving between the Sun and the Earth. Lunar eclipses are caused by the Earth moving between the Sun and the Moon.
- You can experiment with shadows using a torch (the Sun) and a ball (the Moon) to mimic the Full and Half Moon shapes and eclipses. Ask someone to shine the light on the ball. Now walk around and look at the ball from different directions. You should see the different phases of the Moon.